LINDA RUNYON

A SURVIVAL ACRE

50 Worldwide Wild Foods & Medicines

A BASIC GUIDE

LINDA RUNYON'S WILD FOOD COMPANY
101 Train Street, Suite #1
Dorchester, MA 02122
(617) 265-7714 www.wildfood.org

A SURVIVAL ACRE©

Drawings by Linda Runyon

ISBN: 0-918517-03-6

Library of Congress Catalog Card Number: 84-72924

First printing January 1985, Currier Press, Inc.
Second printing June 1985, Nationwide
Third printing March 1987
Fourth printing October 1993
Fifth printing January 1995, 4Mom Ventures
Sixth printing June 1996, Terra Libra II
Seventh printing October 1999, Linda Runyon's Wild Food Company

Health Research
PO Box 850
Pomeroy, WA 99347

www.healthresearchbooks.com

Acknowledgements

My sincerest "thank you" to so many people in and around Indian Lake, New York, without whose support, patience and all-around understanding I may never have experienced my life with Nature in the Adirondacks.

They are: The Gavetts, The Catlins, Ruth Spring, Dorothy and Lynn Galusha, Gert and Dick Frulla, Nancy and Pete Hutchins, Diane and Frank Benton with special thanks to Joe, Marge (former) and Paul Burgess, (Former) Mary Husson, Alice and Walt Sherman, Edward, Janet, John, Walt Jr., Tina and Rick Bennett, Edward Hall, Paul Blanchard, Editor (and friend) Pam Gang, Consultant Lynn Waldron, Survival Acre's Seed Researcher Phyllis Williams, Bookkeeper Gail Goodenough, and Many Others.

Last, but first, to my son Todd, who grew up on my wild concoctions and without whose patience and tolerance this book would not be possible:

To my daughter Kim, who washed her hair with "soap plant" and picked blackberries with the Indian Lake black flies so Mom would have enough to dry for winter:

To her husband Chuck, who ran off the first copies of "Survival Acre" with my grandson Jason's help:

To my oldest son Eric, who climbed down into a 30 foot deep Adirondack well after a pail, and has taught his Mother that nothing is impossible:

To my parents, Ruth and Paul Runyon, without whose love and support and personal appreciation for the Adirondacks I may never have been led to the experiences in this book:

To my former husband Ken Heitz, who tested literally hundreds of "experiments".

To my grandchildren Ceri, Alec, Lee & Jason who tasted the "Wild" with Grandma.

INTRODUCTION

LINDA RUNYON was raised during summers at Nirvana Lodge, a tourist camp owned by her grandparents. As an adult, returning to Indian Lake from the Jersey shore seemed the natural thing to do. Homesteading in the Adirondack Mountains, without modern conveniences, seeking wild foods, cooking outdoors and learning basic Adirondack-Indian ways became the beginning of "A SURVIVAL ACRE".

THE AUTHOR CAUTIONS: please cross reference all plants with several field guides and be positive about your food. Use photo identification. Neither the author nor publisher assumes any responsibility for the individual's interpretation and consumption of wild foods researched in this book.

SURVIVAL ACRE: . . . Remember, window boxes, pails & pots of soil are all perfect ways to grow weeds. Seeing where they are on the Survival Acre will help you determine how to grow them. Gathering your own plants to dry gives you seeds to replant . . . Nature takes over . . . as well as a new appreciation for the world we all share.

On the Survival Acre, any combination of 6 to 10 of your favorite plants would be adequate for any eventuality. Being shown survival by being snowed in, I found the many jars of food a "Godsend". Mixing with water and making a paste, tortillas were the easiest way to nutrition. As an additive to regular foods, the combinations were endless. Stapling a basic 7 to 8 of these vegetables, 3 to 4 teas and a constant nibbling while foraging, hold a basic nutritional value along with fruits and breads. You could begin by identifying 1 or 2 plants in your yard. Check at least three references and start a small filing system. Gather a few jars and labels, then collect and experiment with these nutritious taste treats. Boredom is a word I never use. Everywhere I look the basic word **food** surfaces. Collecting natural food is an enjoyable challenge, and I have found the money savings to be immeasurable.

Jars of dried foods can be a pinch to nutrition: foods can be frozen, boiled, steamed, baked or canned as other vegetables can. In a dry state, these foods will last for years. I have found shelf-life to exceed 5 years, and recently I have found seeds from these stored foods capable of growing successfully with a minimum of care. Although dried foods will brown out when too old, they usually maintain their green nutritious state. Judge by the color, although even when changed they are useful in soups and stews. A family can spend a few hours from season to season and gather a winter supply or enough to supplement as many meals as one chooses with fresh vegetables ... A few minutes for hundreds of "pinches".

There are literally hundreds of plants to choose from when selecting food sources. However, the 50 Survival foods listed here seem to be the most prevalent, and are found naturally throughout the Adirondacks and surrounding states. Once you know what plants you have nearby and which ones you need to transplant into boxes, pails, etc., you could gather free seed at the end of Fall (see seed chart). Clumps of Clover, Milkweed, Aster, Pigweed, Sorrell and Lamb's Quarters are the highest in nutrition and seem to be everywhere in fields and yards and along the roadside.

ALWAYS be selective, leaving spaces between plants, and NEVER pick an area clean. Save all your dandelion leaves **before** you mow your lawn. Picking your foods after a rain is the cleanest method of harvesting. Be sure to cleanse thoroughly before storage or preparation. Shake the leaves dry or lay out in the sun for a few hours . Dried and powdered, these foods will provide you with hundreds of pinches.

As you go along you will see that anywhere there is an area of dirt, there is food. The uses are endless. Perhaps you will experience farming your own yard and eliminate "boredom" from any given minute. Let 10 to 20 feet of grass grow for your field.

HAPPY HUNTING and MAY
NATURE'S FORCE BE WITH YOU!

Miss Linda Runyon

Contents

List of Maps & Illustrations

AND GOD SAID, "*Behold, I have given you every herb bearing seed, which is upon the face of all the earth, and every tree in which is the fruit of the tree yielding seed; to you it shall be for meat. And to every beast of the earth, and to every fowl of the air, and to every thing that creepeth upon the earth, wherein there is life, I have given every green herb for meat: and it was so.*"

Genesis 1:29-30

50 PLANTS, FOODS, TEAS & MEDICINES

1. QUEEN ANNE'S LACE . . . . food
♥2. MILKWEED . . . . food
3. CHICORY . . . . food
♥4. MULLEIN . . . . medicine
5. CURLEY DOCK . . . . food
♥6. THISTLE . . . . medicine, food
7. ROSE HIPS . . . . food
8. BLUEBERRIES . . . . food
9. FIDDLEHEADS . . . . food
10. MINTS . . . . tea
11. WOOD SORRELL . . . . food
12. MUSTARD . . . . food
♥13. PLANTAIN . . . . food
14. CATNIP . . . . medicine
♥15. YARROW . . . . tea
♥16. PINE NEEDLE . . . . food
17. BURDOCK . . . . food
18. PEPPERMINT . . . . food
19. PIGWEED . . . . food
♥20. WEEPING WILLOW . . . . medicine
21. RASPBERRY . . . . food
22. BLACKBERRY . . . . food
♥23. FIREWEED . . . . food
24. MEADOWSWEET . . . . tea
25. BLUE ASTER . . . . food
26. VIOLET . . . . food

♥*Most staple for author*

50 PLANTS, FOODS, TEAS and MEDICINES continued

27. BRACKEN . . . . food
28. BALSAM . . . . food
29. WHITE BIRCH . . . . food, drink
♥30. BULRUSHES . . . . food
31. BUR-REED . . . . food
32. WAPATOO . . . . food
33. WILD RICE . . . . food
34. FALSE SOLOMON SEAL. . . . . medicine
35. WILD WILLOW . . . . medicine
♥36. CATTAIL . . . . food
37. TANSY . . . . medicine
38. HEAL-ALL . . . . medicine
39. STRAWBERRY . . . . food
♥40. RED CLOVER . . . . food
41. WHITE CLOVER . . . . food
♥42. FIELD SORREL . . . . food
♥43. LAMB'S QUARTERS . . . . food
♥44. CHAMOMILE . . . . tea
♥45. THYME . . . . tea
46. PURSLANE . . . . food
47 .WILD LETTUCE . . . . food
♥48. DANDELION . . . . food
49. WINTERGREEN . . . . tea
♥50. GOLDENROD . . . . tea

♥*Most staple for author*

WHERE TO FIND THE BASIC 50

Areas where the Basic 50 Plants are found in the Adirondack North Country, as based on the Author's studies made in and around Indian Lake, Hamilton County, New York State. The locations are as a rule... may be found as exceptions.

ROADSIDES:

Queen Anne's Lace
Milkweed
Mullein
Curley Dock
Chicory
Yarrow
Meadowsweet
Fireweed

WOODS:

Fiddleheads
Dogtooth Violets (blue also)
Pine Needle
Braken Fern
Blue Aster
White Birch
Balsam
Wood Sorrel

BROOK AREAS OR SWAMPS:

Bullrushes
Bur-reed
Wapatoo-Arrowhead
Wild Rice
False Solomon's Seal
Wild Willow
Cattail
Weeping Willow

LIKELY TO POP UP IN A GARDEN:

Lamb's Quarters
Chamomlle
Thyme
Purslane
Wild Lettuce
Dandelion

FIELDS & LAWNS:

Tansy
Heal-All
Strawberry
Red Clover
White Clover
Field Sorrel
Catnip
Plantain
Mustard

Mint
Blackberries
Raspberries
Peppermint
Burdock
Pigweed
Wintergreen
Goldenrod
Blueberries

Rose Hips
Thistle
Lamb's Quarters
Chamomile
Thyme
Purslane
Wild Lettuce
Dandelion

PLANTS TO AVOID

RULES:

- Always be familiar with all dangerous plants in your area.
- Always positively identify all plants you intend to use for medicine or food.
- Crush, sniff, inspect carefully, taste tiny amounts.
- Cross reference with several books, using pictures and all information given.
- Keep all plants away from all children and pets.
- Teach children to keep all plants away from their mouths. Do not let children suck the nectar from any unknown plant. One may be poisonous, another may not. Honeysuckle is OK. Deadly Nightshade is not.
- Avoid smoke from burning plants, as smoke may irritate the eyes or cause allergic reactions.
- Heating or boiling does not always destroy toxicity.
- Store bulbs and seeds safely away from children and pets.
- Do not collect plants from less than 50 feet of roads or from insecticide-sprayed areas. Wash thoroughly.

DANGEROUS—POISON PLANTS

The following poisons are referred from *COMMON POISONOUS PLANTS* by John M. Kingsbury: Information Bulletin # 104, Division of Biological Sciences, 1, New York State College of Agriculture and Life Sciences, Cornell University, Ithaca, New York.

American False Hellibore
Bloodroot
Bouncing Bet
Box
Black Locust
Black Nightshade
Bracken*
Buttercup
Celadine Poppy
Chokecherry
Christmas Rose
Cocklebur
Corn Cockle
Crocus
Daphne
Deadly Nightshade
Dogbane
Dutchman's Breeches
Eastern Lupine
European Bittersweet
Horse Chestnut
Horsetail
Horse Nettle
Indian Hemp
Indian Tobacco
Jack in the Pulpit
Jimson Weed
Leafy Spurge
Lily of the Valley
Mayapple
Marsh Marigold
Mountain Laurel
Moonshod
Pokeweed
Poison Hemlock
Privet
Purple Cockle
Red Baneberry
Rhubarb Leaf
Skunk Cabbage
Snowdrops
Star of Bethlehem
St. Johnswort
True Solomon's Seal
Water Hemlock
White Baneberry
White Snakeroot
Wild Black Cherry
Yews

** Bracken Fern has been known to poison cattle if eaten in large quantities.*

Poisons

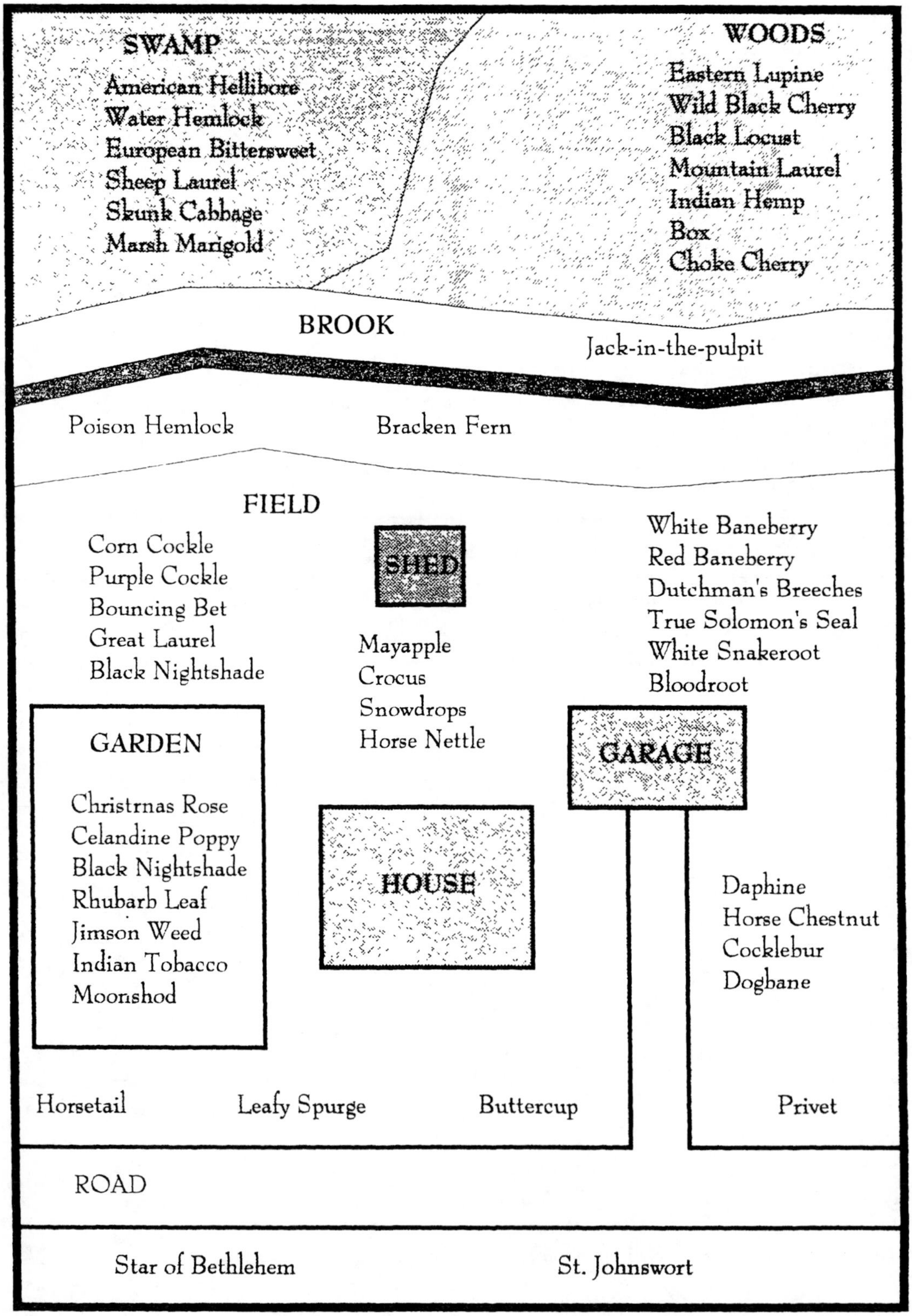

SWAMP

American Hellibore
Water Hemlock
European Bittersweet
Sheep Laurel
Skunk Cabbage
Marsh Marigold

WOODS

Eastern Lupine
Wild Black Cherry
Black Locust
Mountain Laurel
Indian Hemp
Box
Choke Cherry

BROOK

Jack-in-the-pulpit

Poison Hemlock

Bracken Fern

FIELD

Corn Cockle
Purple Cockle
Bouncing Bet
Great Laurel
Black Nightshade

SHED

Mayapple
Crocus
Snowdrops
Horse Nettle

White Baneberry
Red Baneberry
Dutchman's Breeches
True Solomon's Seal
White Snakeroot
Bloodroot

GARDEN

Christmas Rose
Celandine Poppy
Black Nightshade
Rhubarb Leaf
Jimson Weed
Indian Tobacco
Moonshod

HOUSE

GARAGE

Daphine
Horse Chestnut
Cocklebur
Dogbane

Horsetail

Leafy Spurge

Buttercup

Privet

ROAD

Star of Bethlehem

St. Johnswort

GATHERING TIME

ROOTS: Early Spring, before the sap turns.
BARKS: Usually in the Fall.
BERRIES: In season.
SEEDS or FLOWERS: When fully ripened.
MEDICINAL HERBS: Anytime, unless specified.

- Gather the newest plants after a rain, 40 or 50 feet from the side of the road.

- (Adirondack Park) New York State Laws

  TITLE 6 ENVIRONMENTAL CONSERVATION LAW

  190.8 General.

  (G) No person shall deface, remove, destroy or otherwise injure in any manner whatsoever any tree, flower, shrub, fern, moss or other plant, rock, fossil or mineral found or growing on State Land, excepting under permit from the Commissioner of Environmental Conservation and the Assistant Commissioner for State Museum and Science Service, pursuant to section 233 of the Education Law as amended by chapter 121 of the Laws of 1958, nor shall songbirds and their nests and other wildlife be molested or disturbed at any time, except during the open season therefor, if any.

- Gather plants on non-posted access roads or from fields with permission. Pick only when there are several of any one plant. Move from place to place, picking sparingly.

PROPAGATION OF PLANTS

Plants actually thrive on selective picking. The following are plants that improve after being cut back. In some cases, when gathered carefully, the remaining wild foods double or triple in productivity.

*Milkweed
*Mint
*All fruits
*Peppermint
*Weeping Willow Branches
Strawberry Leaves

*Dandelion Leaves
Heal-All Spikes
*Plantain Leaves
Mustard Leaves
Burdock Burrs
*Wood Sorrel

Goldenrod
*Tansy
*Violets
*Pigweed

*Double or Triple

Lewey Lake Campsite, winter watering hole

DRYING OBJECTS

The following list of equipment is handy, if not necessary:

- 2-foot lengths of string
- scissors
- a favorite cutting knife
- gloves
- shovel
- large paper bags
- a pail for live transplants

Use both hands to pick berries. You can pick faster by fastening the top of a plastic bottle around your waist tied with a string.

When drying, you can use clothesline in your garage, attic or kitchen. Pin bundles to the line, or hang them from convenient nails. Never hang them in direct sun, but in warm, airy places. You can dry individual leaves on screens. Berries do well on trays left on the stove pilot or placed high in the kitchen. Place a piece of cheesecloth over the trays to keep out insects.

When the bundles have dried, lay a sheet on the floor. You can easily strip off the leaves and flowers. Scoop them up and store in 1/2 gallon glass jars. 10 minutes of work for a winter supply of nutrition!

Place the jars in darkened areas, such as closets or cellars for long term storage. Use brown glass when it is possible in the kitchen. Decorative storage ideas are endless!

Seeds have been stored by the author and grown successfully after six years of storage.

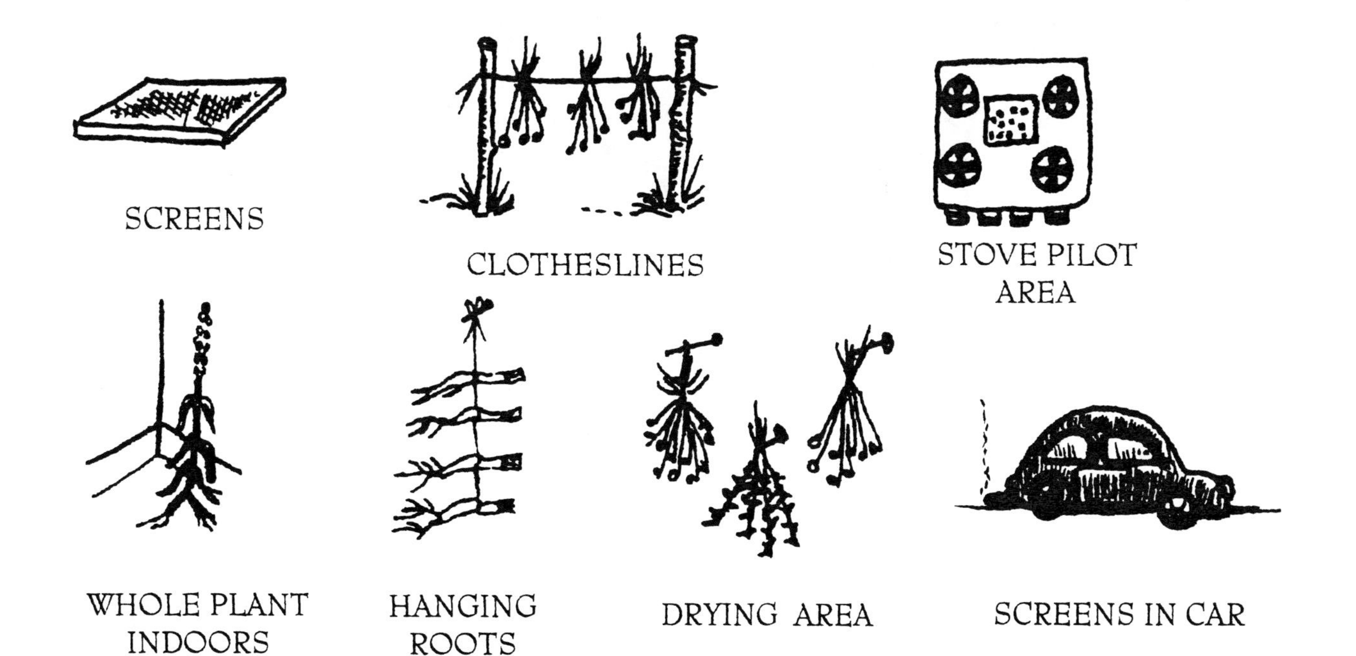

50 Plants, Foods, Teas & Medicines

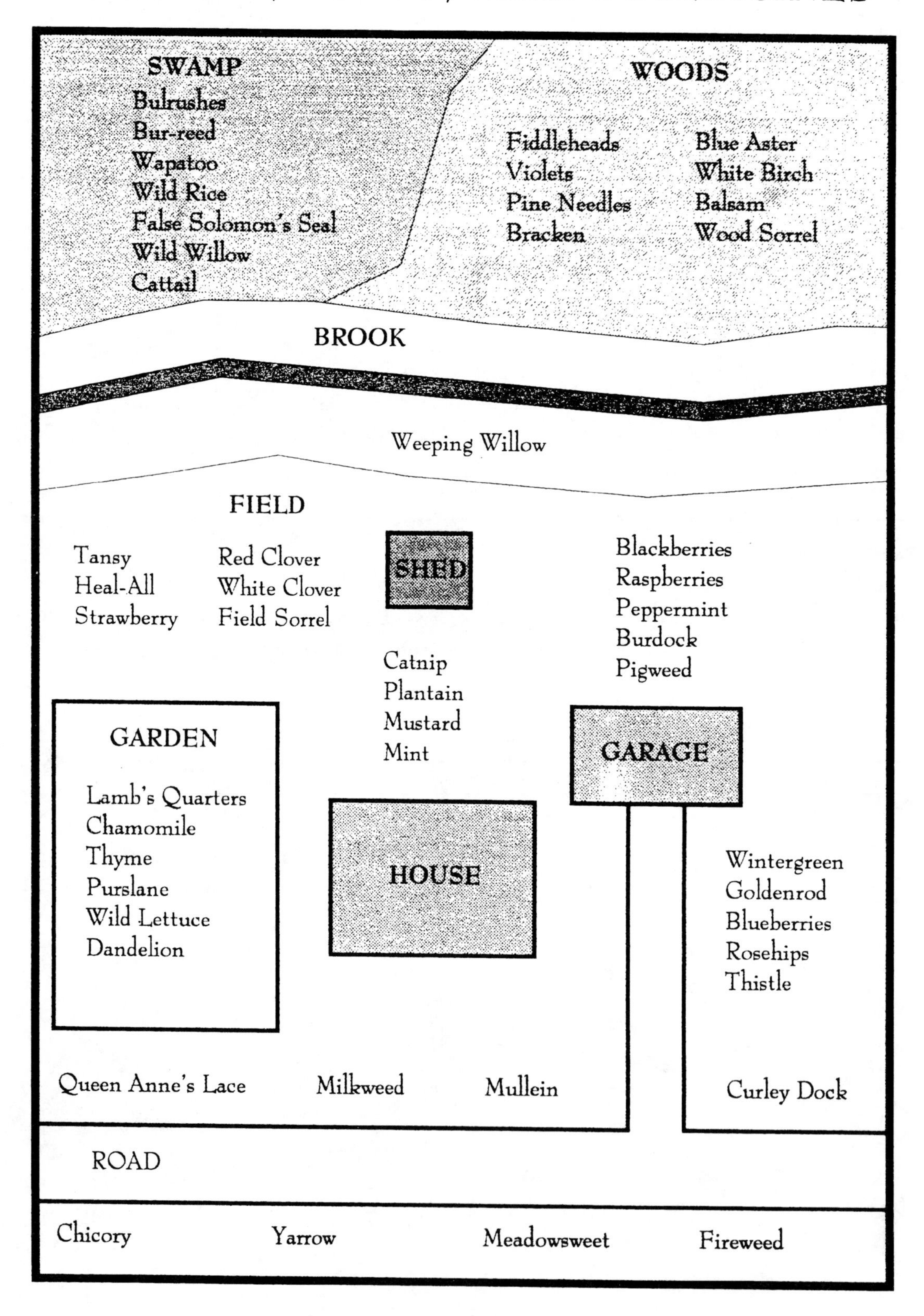

PLANTS, DESCRIPTIONS, USES & SKETCHES

Author cautions: please cross reference each plant with other books and photographs for positive identification.

Statistics and chemical breakdown of Vitamins and Minerals are found *in EAT THE WEEDS* by Charles Harris and *A TREASURY OF AMERICAN HERBS* by Virginia Scully. Sizes of plants based on Adirondack studies.

Descriptions & Uses

QUEEN ANNE'S LACE: *Daucus Carota (L.)* Wild Carrot, Caraway, Bird's Nest, Anise Substitute.
HISTORY:
Indians gathered seeds in early Fall and used them as salt. Wait until head is brownish.
PRESENT:
Bottle the *seeds* for excellent sprouts and use as salt substitute and spice in everyday cooking.
SIZE:
18" to 24" high

MILKWEED: *Ascleptas Syrica (L.)* The taste and Vitamin A content can be compared to fried Okra. (6.3 mgs. per fresh pound); also a broccoli substitute.

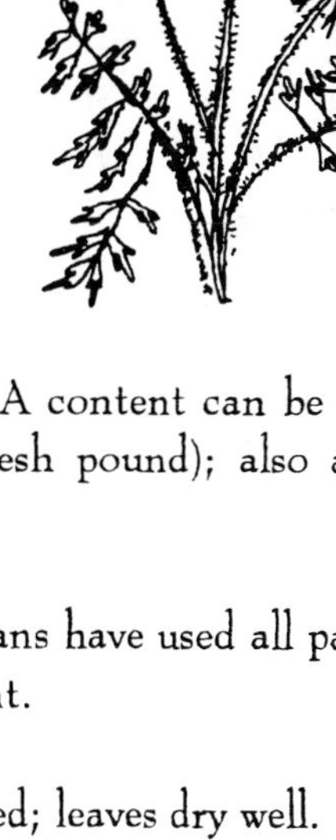

HISTORY:
All Indian tribes used it; Europeans have used all parts of the plant at each stage of development.
PRESENT:
It can be frozen, pickled or canned; leaves dry well.
PARTS USED:

1. Early *shoots* as a vegetable.
2. Ojibwe Indians boiled *heads* or steamed them until soft as with asparagus.
3. Chippewa Indians used the *flowers* either raw or rolled in egg batter and flour and fried.
4. Shoshone Indians boiled the *pods* until they turned green. For personal use, pick the pods before they turn elastic, boil 2 to 3 minutes 2 or 3 times... this removes the slightly bitter milk. Pour out on a bread board, slit lengthwise, salt & dip in flour or oatmeal and then fry until brown.

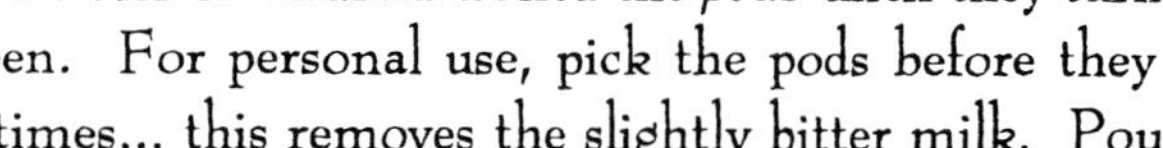

5. Iroquois Indians used all parts. They made gum from the *milk*. To make gum, just pick any part of the plant: it will bleed milk. Gather and roll in the palm of your hand until it is an elastic sweet gum.
SIZE:
24" to 32" high

CHICORY: *Chichorium Intybus (L.)* Use a decoction of ground *roots* or *leaves* as diuretic and laxative. Angina, melancholia, hysteria, neurosis, hepatitis, gall bladder, nephritis.
A staple French food.
PRESENT:
Use *roots* dried and ground; add them half and half to coffee to stretch supply. Use *leaves* raw in salads. A staple green vegetable in many countries.
SIZE:
24" high

MULLEIN: *Verbascum Thaspus (L.)*
I call it the *Aloe Vera of the North.*
see Medicines.
SIZE :
36" to 40" high

CURLED DOCK: *Rumex Crispus (L.)* Yellow Dock, Garden Patience.
HISTORY: All Tribes, Europe.
USES: Roots (See Medicines).
LEAVES:
Cook young for 10 minutes, butter as vegetable. Freeze, steaming 1 minute.
SEEDS:
When brownish, and peel up off stem easily, collect and grind for flour...very nutritious in A! Use sparingly, can easily over-do vitamin A! Used after winter's browning, dried flower material for arrangements.
ROOTS:
See Medicines. Used as a yellow dye.
SIZE: 14" to 24" high

THISTLE: *Sonchus Oleraceus (L).* Sow Thistle.
HISTORY:
Indians ate as salad, herb as medicine.
Pioneers scattered English thistle.
PARTS USED:
Early *shoots, roots, stalks.* Salads, green boil spiny *leaves,* freeze as vegetable. *Root, stalks,* boil or fry.
Nourishing and restorative.
SIZE.
18" to 30" high

ROSE HIPS: *Rosa Specie (L.)* Fruit... contains 60X more Vitamin C than a lemon and 40X the C of an orange.
PRESENT:
Use the *hip part* of any Rose, dry for tea, store in glass container. Winter hips a C tea.
SIZE: 14" to 30" bush

BLUEBERRIES: *Vaccinam Myrtillus (L.)*Fruit.
HISTORY:
All Tribes.
.PRESENT:
Cultivated as a fruit in all countries.
USES:
Dry fruit on trays in warm airy areas. Good to dry over the pilot light on top of the stove. They rattle when dry.
. . see Medicines.
Size:: 14" to 20" bush.

FIDDLEHEADS: *Pteridium Aquilinum (L.)*
HISTORY:
All tribes.
PRESENT:
Gourmet food and vegetable in restaurants.
PARTS USED:
Curled *fronds.* Remove the white fur under running water.
USES:
Eat with meat, fish and other vegetables. Freezes nicely. Boil twice, 5 minutes each, and serve with oil and vinegar.
SIZE: 12" to 24" high

MINT: *Mentha Arvensis (L.)*
Many varieties in North.
HISTORY:
All tribes.
PRESENT:
Staple medication of the author. Also a summer drink, either cold or hot. Make a mint vinegar by placing the *leaves* in a jar of vinegar. See Herbal Vinegar recipe.
SIZE:
10" to 14" high

WOOD SORREL: *Oxalis Acetosella (L.)*
The "Shamrock", trefoil.
HISTORY:
All tribes.
PRESENT:
5 to 15 leaves an excellent source of Vitamin C. An instant thirst quencher while out walking in the woods, tangy taste.
USES:
Dug whole as a Vitamin C source, pinch into soups and stews.
SIZE: 2" to 6" high

WILD MUSTARD: *Barbarea Verna and Vulgaris (L.)*
Wintercress.
HISTORY: All tribes.
PRESENT: A vegetable.
USES:
Freeze, steam or dry *leaves* and *seeds* and pinch into soups, etc. Use seeds as everyday spice.
SIZE:
10" to 14" high

PLANTAIN: *Plantago Major (L.)*
PRESENT:
Easy to run fingers up the spikes and collect the *seeds* for winter storage; sprout and eat.
USES:
Leaves used as turnip green substitute. Freeze or dry and put pinches in casseroles, etc.
SIZE: 1" to 2" high

Long Leaf

CATNIP: *Nepeta Cataria (L.)*
HISTORY:
Medicinal herb.
PRESENT:
Cultivated for commercial use. See Medicine. Use as you would Mint for tea, etc.
SIZE:
12" to 16" high

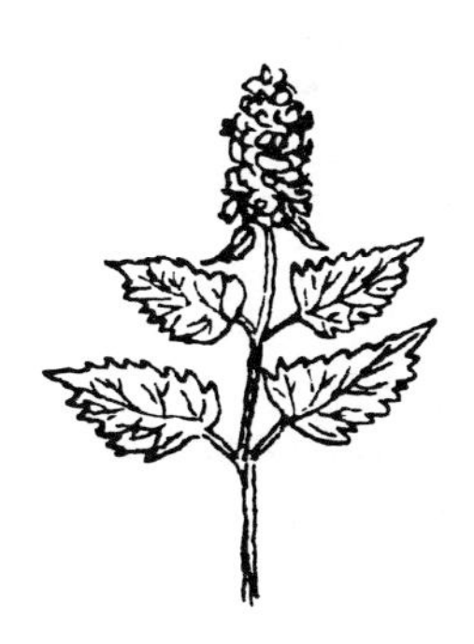

YARROW: *Achillea Millefolium (L.)*
"Chipmunk's Tail".
HISTORY:
All tribes.
PRESENT
Raw, stimulant, high in Vitamins A and C. Dry and use as a staple tea.
USES:
Leaves. flowers as a dentifrice. Scrub the teeth with the leaves, raw; also an astringent.
SIZE:
16" to 24" high

PINE NEEDLE: *Pinus Strobus (L.)*
Very high in Vitamin C.
HISTORY:
Used as a thirst quencher by all tribes.
PRESENT:
Take it on a trip as a taste treat; chew one needle slowly for a taste of the woods.
SIZE: Length needle 2" to 4" long

BURDOCK BURRS: *Arctium Lappa (L.)*
Burrs, Beggars Buttons.
HISTORY:
All tribes.
PRESENT:
Craft material. Let dry on bush naturally, pick and press into baskets, creatures, etc., then varnish. The *burr* is also a source from which insulin is made. The *leaves* are used the same way as any Spring green. When the *burrs* are green on the bush, shell and eat as you would any raw vegetable.
SIZE: 36" to 40" bush

PEPPERMINT: *Mentha Piperita (L.)*
The common field type, lavender fuzz tops.
HISTORY:
All tribes.
USES:
Leaves and *flowers* as a digestive aid or tea. Use powder for delicious tea. Stores well.
SIZE: 12" to 16"

Eastern

PIGWEED: *Amaranthus Blitum (L.)* Amaranth, an asparagus substitute.
HISTORY:
Came from Mexico; staple food of the Zapotec Indians.
PRESENT:
Staple food, cultivated by the Navahos.
USES:
Freezes well as a staple vegetable; complete lettuce substitute. Dry and powder *leaves* and put a pinch into everything. The sweet, raw *leaves* and black *seeds,* found in the Fall, can be pounded to a fine black powder and used as a flour. Author suspects Pigweed and Lambs Quarters may be the same plant but male and female. Always seem to grow side by side.
SIZE: 14" to 24"

Western

WEEPING WILLOW: *Salix (L.)*
HISTORY:
Aspirin derivative, charcoal, crafts.
PRESENT:
Acetasalacyllic acid; aspirin. A rose to tan colored dye; baskets; whistles, fishing poles.
USES:
Make infusion of *bark* and *twigs*. *Leaves* and *catkins* may be used as aphrodisiac.
SIZE: Tree

RASPBERRIES: *Rubus Series (L.)* Fruit.
HISTORY:
All tribes.
PRESENT: Cultivated.
Preserves, jelly, canned or frozen berries. Boil leaves for medicine. See Medicines.
SIZE: Trailing bush

BLACKBERRIES: *Rubus Series (L.)* Fruit.
HISTORY: All tribes.
PRESENT: Cultivated.
USES:
Jellies, syrups. Cans and freezes well. Vitamin C source.
See Medicines.
SIZE:
Trailing bush.

FIREWEED: *Epilobium Angustifolium (L.)*
Willow herb, found in fire waste areas.
HISTORY:
Staple food in the East and Far East.
PRESENT: Asparagus substitute.
USES:
Tips, before they flower, freeze well as vegetables, or eat raw in salad; *buds* can be eaten raw, stewed, or put in casseroles. The *pith* of the stem can be used in soups and stews, and the *dried leaves* used as a basis for teas. The cotton-like *fluff* found in the Fall makes good tinder, and is an easy way to plan where you will pick next year.
SIZE: 24" to 36" high

MEADOWSWEET: *Filipendula Ulmaria (L.)* Herb.
HISTORY:
An invigorating drink; staple tea of Pioneers.
PRESENT:
Staple tea: dry *flowers* and *leaves* for a winter supply. Powdered sugar substitute.
USES: Dry *flowers* and store; use as above.
SIZE: Bush.

BLUE ASTER: *Aster Nemoralis (L.)* 250 species.-
blue, purple, and white in Adirondacks.
HISTORY: Staple food of the Iroquois Indians.
PRESENT: A storable, staple food, high in nutrition.
USES: *Tips* before they flower and green *leaves* as greens in salads. Dry plant by hanging upside down, strip flowers and leaves. Powders easily for use as seasoning. Pinch into cooking. Especially good in lasagna.
SIZE: 24" to 30" high

VIOLETS: *Viola Species (L.)*
Source of Vitamin C.
HISTORY:
All tribes. Europeans use for candy.
PRESENT:
Dried or candy, using *flowers* or *leaves*. Freeze in ice cubes for a decorative touch.
SIZE: 4" to 8" high

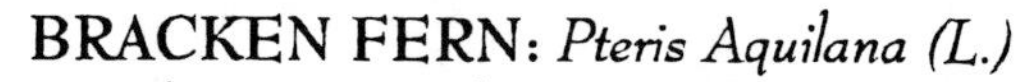

BRACKEN FERN: *Pteris Aquilana (L.)*
Fronds, Sweet Brake.
HISTORY:
Far East to West, to the Eastern US. Oldest collected vegetable in history.
PRESENT:
Eaten raw in the woods, it changes your perspiration to a woodsy smell while hunting.
USES:
Freeze, or boil and serve as a vegetable with butter. Delicious pickled in your favorite recipe. A greenish yellow to gray dye is produced from the uncurled *fronds* in early Spring. Caution: may be slightly toxic in large amounts.
SIZE: 16" to 24" high

BALSAM: *Balm of Gilead (L.)*
Christmas tree.
HISTORY:
Indians chewed while in the woods as a thirst quencher.
PRESENT:
Same as above; also an excellent source of Vitamin C.
SIZE: Tree.

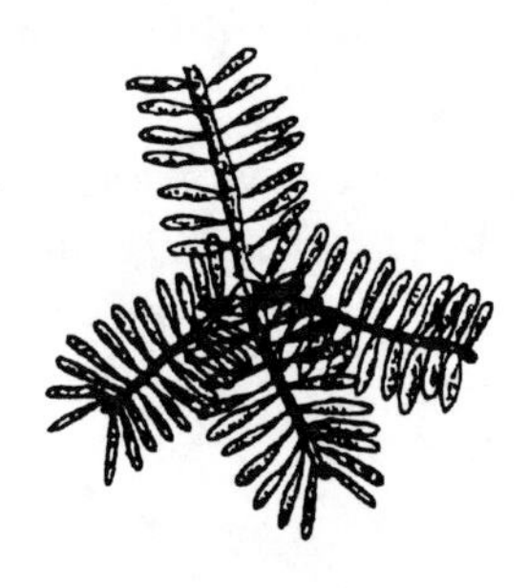

WHITE BIRCH: *Betula Alba (L.)*
Paper Birch, grind inner bark.
HISTORY:
Scandinavians boil, bake and add sawdust for use in baking as flour. Indians used for tea and for utensils, etc.
USES:
As above and for canoes, pots, pails, crafts.
SIZE: Tree.

BULRUSHES: *Scirpus Validus (L.)*
Adirondack Great Bulrush.
HISTORY:
As a food throughout history from Moses to all Native American tribes.
PRESENT:
Roots can be ground into white flour and is stored easily. *Tops* also, before they turn to fluff (gone to seed).
SIZE: 24" to 36" high

BUR-REED: *Sparganium Androcladum (L.)*
Found in swamps.
HISTORY:
Food for most tribes; medicine.
USES:
Stems and *burrs* can be peeled and cut up for use as a vegetable. "The seed of the burr, when drunk in wine, would cure a snakebite." *Dioscordes,* 50 years BC
SIZE.: 10" to 16" high

ARROWHEAD: *Sagittaria Latifolia (L.)*
Wapatoo, Wild Potato.
HISTORY:
"Can be ground fine, a flair for pudding, cakes, etc. They are nearly equal to Irish Potatoes and are a bread substitute." Quoted from Lewis and Clark.
USES:
Dig with shovel carefully, or with a pitch fork if you have one. *Potatoes* fall off of the roots easily. String with coarse thread and hang to dry in a warm, airy place. Can be a staple potato, cooked all winter, or even be ground into a fine flour.
SIZE: 12" to 16" high

WILD RICE: *Zizania Palustris (L.)*
Adirondack Variety, many types.
HISTORY:
All tribes; stripped by hand into canoe.
PRESENT:
Gathers easily; store in glass containers; boil or fry as any rice. Use as sprout or as stuffing. It is smaller than regular rice and needs no separation. Fry in butter or olive oil.
SIZE: 18" to 24" high

FALSE SOLOMON'S SEAL: *Smilacina Racemosa (L.)*
Positively identify from the *True Solomon Seal* . . . POISON. See photo in reference.
HISTORY:
Roots were dug by Algonquins and Iroquois. A source of Vitamin K, useful in clotting the blood. An old Indian method. The Indians made necklaces from the roots which were then pounded with a rock and used when needed.
SIZE: 18" to 24" high

WILLOW: *Salix Nigra, Alba (L.)*
Wild; See Medicines.
SIZE: Tree.

CATTAIL: *Typha Latifolia (L.)*
"The Supermarket of the Swamp", all countries.
HISTORY:
Used throughout history as a food.
PRESENT:
Cultivated - farming yields 140 tons per acre, raw, or 32 tons of pure white flour!
USES:
Collect *root* stalks in early Spring. Dig, dry, peel and pound into flour. Collect *shoots* in early Spring for use as a vegetable. *Leaves* and *stalks* can be woven into seats, mats, chairs, and bags or baskets. *Heads* can be eaten as corn-on-the-cob; see Recipe. *Fluff* may be used as pillow stuffing. 20 to 30 *pollen heads* yield a loaf of bread. Mix half and half with whole wheat flour. Contains sulfur, protein, phosphorous, carbohydrates, & vitamins. See Medicines.
SIZE: 3' to 4'

TANSY: *Tanacetum Vulgare (L.)*
Bitterbuttons, pepper substitute.
HISTORY: Herb and preservative . . . not a food!
PRESENT:
Rub on meat and fish as a preservative. Add to cooking as an herb.
USES:
Collect the *leaves* in early Spring. The whole plant dries well in bunches. Cut off *buttons* and store as seeds in Fall.
SIZE: 24" to 30" high

HEAL-ALL: *Prunella Vulgaris (L.)*
HISTORY: All tribes.
USES:
To treat diarrhea or as a throat gargle.
SIZE:
6" to 12" high

STRAWBERRY: *Fragaria Virginiana (L.)*
HISTORY: All tribes.
PRESENT:
Vitamin C source. Jams, jellies or preserves; cans and freezes well. Use *leaves,* raw or powdered, as herb, pinched into other dishes. Raw, serves as a dentifrice; removes tartar. Indians called it the heart berry.
SIZE: 2" to 8" high

RED CLOVER: *Trifolium Pratense (L.)*
HISTORY: All tribes.
PRESENT:
Protein . .. a staple vegetable food in China.
USES:
Dry *tops* for tea. Use *whole plant* raw, chopped, frozen or canned. Powder for salt substitute. Wonderful vegetable protein; serve any style.
SIZE: 10" to 14" high

WHITE CLOVER: *Melliotus Alba (L.)*
HISTORY:
All tribes.
PRESENT:
Eat raw or dry, making tea from the *heads.* Attractive in your favorite jar. Also a moth ball substitute – dried *heads.*
SIZE:

2" to 4" lawn

FIELD SORREL: *Rumex Acetosella (L.)*
PRESENT:
Use a pinch in all dishes as a salt substitute. Gather and bundle; dry and strip off *seeds* and *leaves.* Store in glass container. Makes excellent Vitamin C soup.
SIZE: 6" to 12" high

LAMBS QUARTERS: *Chenopodium Album (L.)*
Goosefoot. Most prolific and nutritious of all garden weeds.
HISTORY:
All tribes; cultivated by Navahos.
PRESENT:
Whole plant used as vegetable, in salads, or boiled. You can quickly gather *seeds* for winter supply of flour, vegetable or for sprouts. Cook as mush, freeze, can or pickle. To make your own sprouts, gather seeds, wash, drain and rinse twice a day for 2 days until they sprout. Use as desired. Add half and half to regular flour; a very nutritious flour. Dry plant in bundle on clothesline, strip off and store in glass jars.
SIZE:
18" to 24"

CHAMOMILE: *Anthemis Nobilis (L.)*
PRESENT:
Used throughout Europe. Cultivated for commercial use. Broccoli substitute. Also used as a tranquilizer tea.
USES:
Fresh, boil *heads* and *whole plant* and eat as a vegetable or as a soothing tea.
Collect *heads* when mature for fine tea.
SIZE:
4" to 10" high

THYME: *Lepidium Virginicum (L.)* Peppergrass.
HISTORY:
All tribes used as stimulant. Egyptians used in embalming.
PRESENT:
Use the *whole plant,* particularly the leaves and flowers, as a spice in cooking and on meat and fish. Freezes well. Leaves will keep 1 week or more in the refrigerator, take as you need. Chew raw or use the same way as commercial spice on eggs, potato salad. etc.
Dry on screen and powder. A staple stimulant tea for "hangovers". See Medicines. 1 steeped cup will do the trick.
SIZE: 4" to 6" high

PURSLANE: *Portulaca Oleracea (L.)*
Beds itself into a carpet in rich soil.
HISTORY:
Staple food in Mexico.
PRESENT:
Cultivated in Europe and Asia.
USES:
All parts; steam, chop, pickle, eat raw or dry and powder for herb. Freezes well.
SIZE: 1" to 2" high

WILD LETTUCE: *Lactuca Scariola (L.)*
Prickly Lettuce, Opium Lettuce.
HISTORY:
Extensively used by all tribes. Lettuce cultivated from this plant.
USES:
Early *shoots, leaves* are steamed. Leaves freeze as a vegetable.
SIZE: 24" to 36" high

DANDELION: *Taraxacum Oficinale (L.)*
Spinach substitute. High in vitamins.
HISTORY:
Staple food of all cultures, as 300 BC in Arabia and Egypt.
PRESENT:
Cultivated by farmers. Seeds now sold for planting.
USES:
Whole plant: dig *roots* 2 to 3 inches under soil. slice and eat raw. *Leaves,* boil, freeze, can or dry and powder for pinches, or for an invigorating tea. As an ointment, a cure for eczema. See Medicines.
SIZE: 2" to 8" high

	A pound of DANDELION		A pound of LETTUCE	
Protein	12 3	gms	3 8	gms
Fat	3 2	gms	0 6	gms
Carbohydrate	40.0	gms	9 1	gms
Calcium	849	mgs	194	mgs
Phosphorous	318	mgs	63	mgs
Vitamin A	61,970	units	5060	units
Iron	14	mgs	14	mgs
Thiamin	85	mgs	26	mgs
Niacin	3.8	mgs	6	mgs
Vitamin C	163	units	6	units

WINTERGREEN: *Gautheria Procumbens (L.)*
Partridge berry, teaberry, boxberry.
HISTORY:
All tribes.
PRESENT:
Cultivated for flavoring.
USES:
Dry for a stimulant tea. Pick and chew while walking in the woods. Choose selectively from a large patch. Transplants easily and roots quickly. Loves softwood, pine. Grows well in acid soil.
SIZE.
2" to 6" high

GOLDENROD: *Solidago Odora (L.)*
Anise substitute.
HISTORY:
Pioneers built their wagon rings around the fields of goldenrod and called it
their "Golden Elixir."
PRESENT:
An effective blood tonic tea (pick-me-up). Dry in bundles. Strip leaves for infusions. Easy to store yellow fluff for use as an aromatic tea. Leaves medicinal also. See Medicines.
SIZE :
24" to 36" high

TIPS

PINCH TO NUTRITION

Foods highest in nutritional value can be used in powdered form. Keep in your favorite jars ...pinch in everything. Most vitamins stay intact for years. A PINCH TO HEALTH.

HIGHEST IN NUTRITION:

Lamb's Quarters
Thistle
Pigweed
Plantain
Blue Aster
Dandelion

HIGHEST VITAMIN C SOURCE:

Rose Hips (60x a lemon)
Wood Sorrel
Bracken Fern...raw fronds
Thyme
Yarrow
Pine Needles

PICKLES:

Best results are obtained by using individually, but any combination can be used.

Milkweed Heads
Cattail
Fiddlehead Fronds
Burdock
Purslane
Bracken Fern

MOST FREEZABLE:

Cattails, inner shoots	3 min.
Milkweed Heads	3 min.
Fiddlehead Fronds	7 min.
Purslane	2 min.
Lamb's Quarters	3 min.
Chicory	3 min.
Dandelion	3 min.
Blueberries	1-2 min.
Thistle	1-2 min.
Pigweed	1-2 min.
Mustard	1-2 min.
Goldenrod	1-2 min.
Red Clover	1-2 min.
Mint	1-2 min.
Aster	1-2 min.
Wapatoo	1-2 min.
Plantain	1-2 min.

STAPLE TEAS:

Strawberry
Raspberry
Dandelion
Rose Hips
Mullein
Goldenrod
Yarrow
Mint
Thyme

SPROUTS:

Lamb's Quarters, seeds
Queen Anne's Lace, seeds
Clover, heads
Dandelion, seed heads
Thyme, seed spikes
Burdock, inner core of burr
Plantain Spikes, easiest

Teas

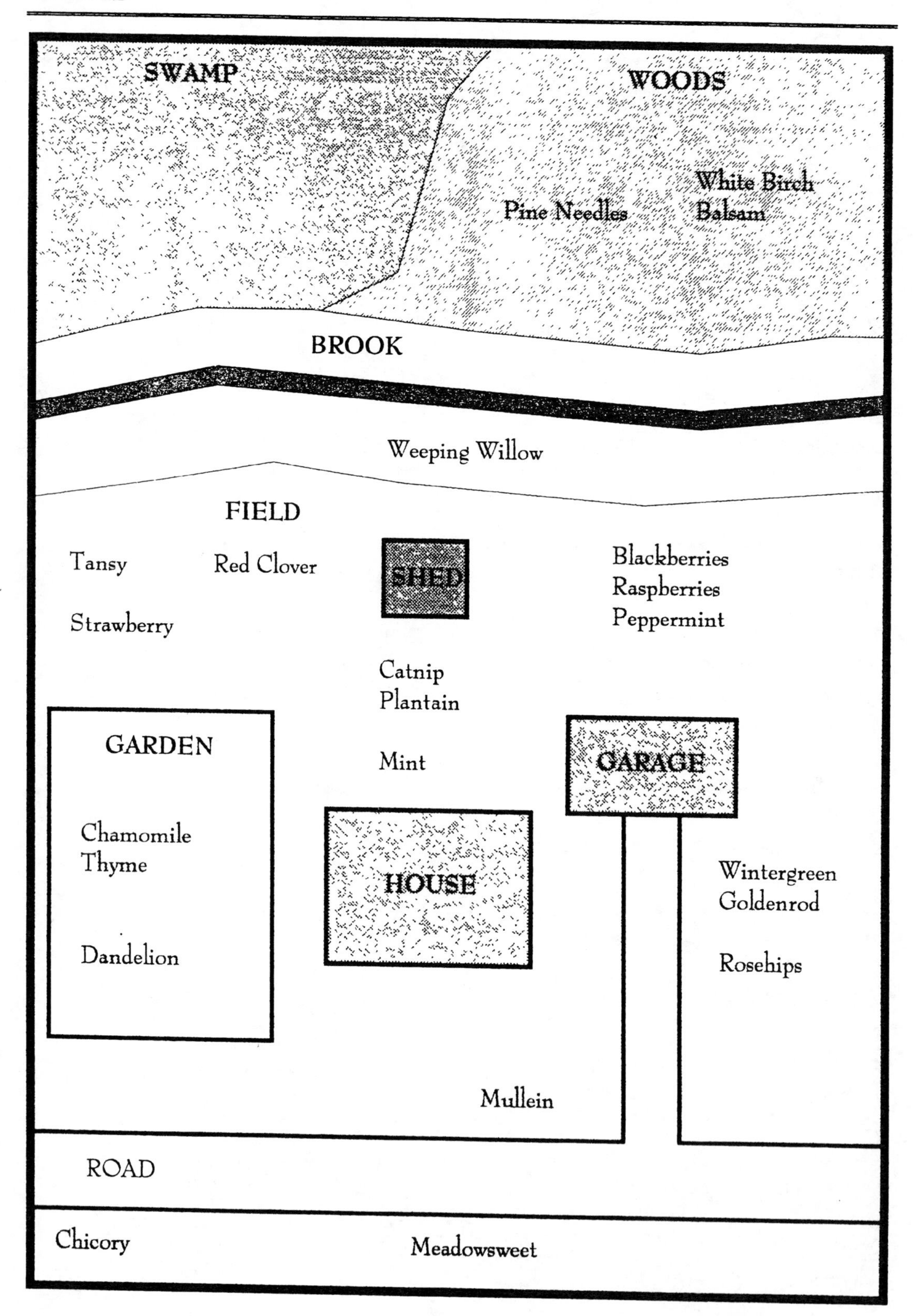
SWAMP
WOODS
White Birch
Pine Needles
Balsam
BROOK
Weeping Willow
FIELD
Tansy
Red Clover
SHED
Blackberries
Raspberries
Peppermint
Strawberry
Catnip
Plantain
GARDEN
Mint
GARAGE
Chamomile
Thyme
HOUSE
Wintergreen
Goldenrod
Dandelion
Rosehips
Mullein
ROAD
Chicory
Meadowsweet

RECIPES

CATTAIL POLLEN PANCAKES (Sunshine Flapjacks)

Collect *cattails* when they have turned brown, after they have almost shed their pollen. Exact time varies, as the cycles complete themselves at different times in each swamp. Shake the pollen into a large bowl. Stores well in glass. Batter for 2.

Ingredients:

Pollen from 1 large cattail
1/2 cup unbleached flour
Your favorite pancake recipe

CATTAIL CORN

Collect the *heads* of the cattails after the spikes have turned green, in early June or so, before the pollen appears. Cut off and boil the heads as you would for corn on the cob, 7-10 minutes. Butter with salt and pepper, or salt with Queen Anne's Lace or powdered Clover as salt. Use Wild Thyme as pepper. Highly nutritious meal!

CATTAIL PICKLES

Choose cattails with the largest stems, 1/2 to 1 inch in diameter (at the height of the summer). Cut cattail at water level. Peel the stem, finding an inner white core. Lay this core on a board, and cut to pickle size (5" to 6" for standard jar). Treat as a cucumber, process the same way, using your favorite pickle recipe. Delicious!

MILKWEED CASSEROLE

Snap off broccoli-like heads before they flower, including the leaves nearest the head. Wash thoroughly under running water. (Serves 4).

Ingredients:

1 quart heads
2 cups milk
bread crumbs

Pour washed heads into pre-greased baking dish. Cover with milk. Sprinkle bread crumbs on top. Cook at 350 degrees for 20-30 minutes. Serve.

SORREL SOUP

Either Wood Sorrel or Field Sorrel.

Ingredients:

2 tablespoons butter
2 tablespoons flour
4 cups Sorrel leaves
1 chopped onion
4 cups milk

Take 4 cups leaves, wash well. Place in a pot and cover with water. Simmer slowly for 1/2 hour. Blend in one chopped onion, butter, flour and 4 cups milk. Simmer 10-15 minutes and strain. Serve.

INSTANT RASPBERRY WINE

Ingredients:

4 quarts ripe raspberries | 4 pounds sugar | 2 quarts sherry

Chop berries in a large bowl, until juicy. Stir in 4 pounds of sugar. Mixture should stand for 3 days. Cover loosely. Stir daily. Pour off clear liquid and add 3 quarts Sherry for each quart juice. Blend well. Enjoy!

CLOVER CASSEROLE

Ingredients:

2 cups favorite tomato sauce
1 teaspoon Queen Anne's Lace
1 teaspoon thyme
4 cups washed Clover, chopped fine

Steam clover for 10 minutes. Add to greased baking dish with tomato sauce and spices. Bake at 350 degrees for 20 minutes.

HOT CLOVER & RICE

A protein delight! Serves 3. Ref.: Randy Leone

Ingredients:

1 cup milk or butter sauce
4 cups fluffy cooked rice
2 cups washed Clover leaves

Add rice to greased baking dish. Stir in clover and milk. Serve hot!

HERBAL VINEGAR

This recipe in for 8 pints of herbal vinegar, or 8 gifts?

Ingredients: (Any 6 in combination.) Do cut the leaves or flowers of the following:

Queen Anne's Lace
Strawberry Leaves
Blackberry Leaves
Violets
Lamb Quarters
Pigweed
Thyme
Mint
Wintergreen
Catnip
Tansy

glass 1/2 gallon jug
1/2 gallon vinegar

Cut leaves or flowers of 6 weeds. Add leaves and flowers to glass jug after washing. Fill glass jug 1/2 to 2/3 full of leaves. Add regular 1/2 gallon of cider vinegar, stir. Shake every day for at least 3 weeks. The longer it sets, the more potent. Put high on a shelf, the warmer the better. Pour into pints, label and ENJOY!

LAMB'S QUARTER FLOUR

Dry a bundle of lamb's quarters upside down in a warm area. Bundle dries in 3-4 days. Lay a clean sheet on the floor when dry and strip leaves/seeds from stems. Put the mixture through a grinder...INSTANT nutritious flour to be used in the same way as regular flour.

You can make flour from: Plantain, Clover, Strawberry Leaf, Pigweed and Dandelion.

Runyon

Vegetables & Salad Greens

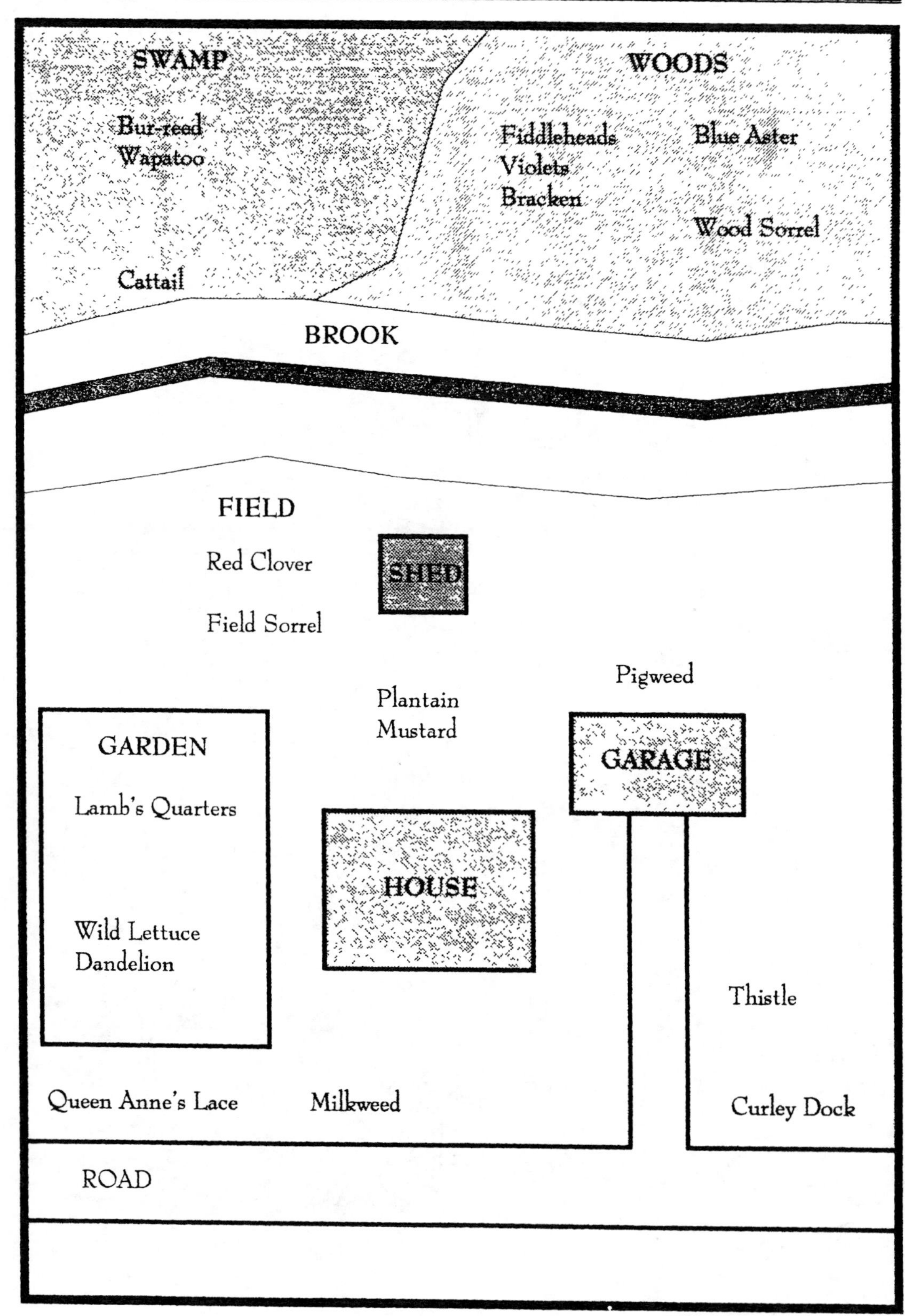

MEDICINES

The American Indian used nearly all of the Basic 50 Survival Plants for their daily needs. Eaten as vegetables every day, the medicinal needs were naturally satisfied. The facts noted here are gathered from several references. I consider the opposite to be basic medicines, and they are staples in my medicine cabinet. Found in my yard, dried and powdered, as a rule are blended 1 OUNCE to 1 PINT of water...2 CUPS boiled to a medicinal smelling strength taken in 1 OUNCE doses 3 or 4 times a day, seemed sufficient. None of these plants are dangerous if taken this way and the dosage may be varied according to the need.

STIMULANT	Yarrow	BLOOD TONIC	Goldenrod
TRANQUILIZER	Chicory, Mullein, Chamomile	BLOOD COAGULANT	False Solomon's Seal
STOMACH soother	Mint	VITAMIN C source	Rose Hips
BURN medicine	Mullein, Yarrow	ANTISEPTIC	Plantain, Mullein, Yarrow
COUGH medicine	Mullein		
ASPIRIN substitute	Wild or Weeping Willow	SORE THROAT	Heal All

METHODS OF PREPARATION

DECOCTION:

Cold water added to plant mixture; 1 oz. to 1 pt. water and simmered gently, with a lid, 1/2 hr. Cooled, then strained.

SYRUPS:

Add 1 oz. of glycerin to substance and can as you would any fruit or berry. STORE OUT OF LIGHT IN COLORED GLASS JARS.

FERMENTATION:

Dip cloths or heavy towels into infusion, wring out and apply to wounds.

INFUSION:

Prepare by pouring boiling water over plant mixture.

OINTMENTS:

1 oz. powdered material to 2 oz. Vaseline.

ESSENCE:

1 or 2 oz. essential oil per 1 qt. Grain alcohol.

TINCTURES:

Spirits of wine can be used for drugs containing gummy or resinous material, or in any situation where extraction by alcohol rather than H_2O is needed...2 oz. bruised to 1 qt. water.

LIQUID EXTRACT:

Concentrated plant values by evaporation. Store extraction, thin later by adding H_2O.

PLASTER:

Bruise leaves, roots, etc. Apply to surface between 2 pieces of cloth.

MEDICINES

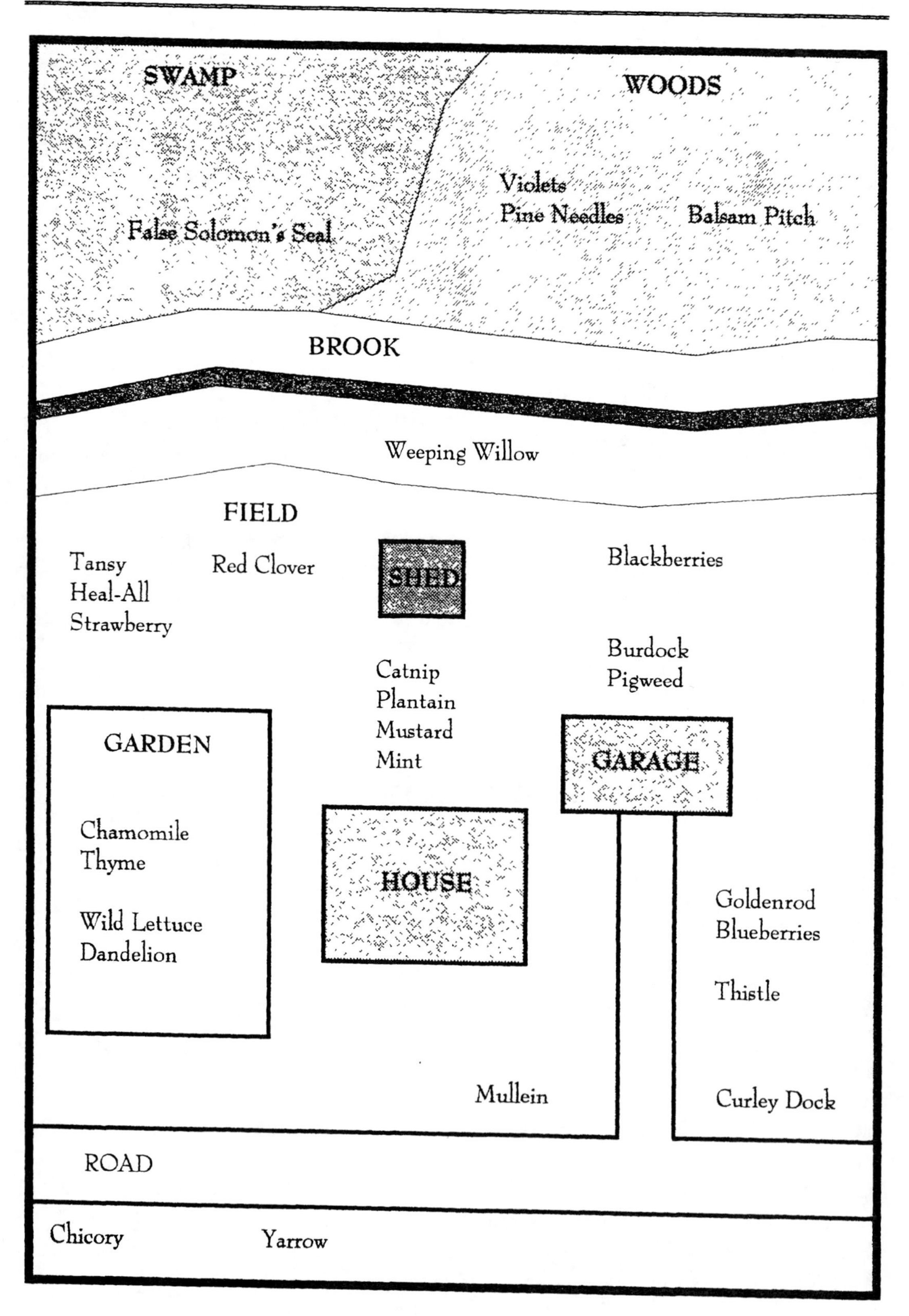
SWAMP
WOODS
Violets
Pine Needles
Balsam Pitch
False Solomon's Seal
BROOK
Weeping Willow
FIELD
Tansy
Heal-All
Strawberry
Red Clover
SHED
Blackberries
Burdock
Pigweed
Catnip
Plantain
Mustard
Mint
GARDEN
Chamomile
Thyme
Wild Lettuce
Dandelion
HOUSE
GARAGE
Goldenrod
Blueberries
Thistle
Mullein
Curley Dock
ROAD
Chicory
Yarrow

MEDICINES

The medical uses listed here are traditional Indian ones. I have chosen only the most common methods, and have used the most easily obtained plants... See SURVIVAL MAP ...Powder all for storage or preparation. See Latin names under foods. Cross reference before use.

CHICORY: Use a decoction of ground *roots* or *leaves* as diuretic and laxative. Angina, melancholia, hysteria, neuroses, hepatitis, gall bladder, nephritis.

MULLEIN: Diuretic, anodyne, anti-spasmodic, tranquilizer, antiseptic.

1. Steep *leaves* and inhale steam for sore throat or asthma relief.
2. Steep one ounce of *leaves* in pint of boiling water and cool. Drink cold to cure diarrhea.
3. Smoke *leaves* for lung trouble.
4. EXTERNAL USES: Ointment soothes wounds, inflamed tissue, burns, fractures (swellings), sprains.

HINT: I take one *whole leaf* and pound with a rock, wrapping it around affected part. This will also draw out splinters and stop bee sting discomfort. Boil and use concentrated as a skin softener.

CURLED DOCK: Listed by the Department of Agriculture as medicine. *Roots:* collect in late summer, split while green, dry and grind for skin infections, itching, irruptions. Collect green and use as a poultice for wounds.

THISTLE: As a tea, slightly relaxing effect. Nourishing and restorative. Decoct *roots* for diarrhea; mashed for aching ear; juice in mouth for toothache; powdered *seed* mashed for diuretic, burns; a decoction of flowers for relief of gonorrhea.

BLUEBERRY: Make syrup or dry until they rattle; store in glass jars. 4 or 6 *berries* will cure diarrhea in most cases. Contains the chemical myrtillian. Chew very slowly & thoroughly for best results.

MINT: Use *whole plant*; infusion or fermentation. Steep in tea for stomach problems, diarrhea, or head cold relief.

MUSTARD SEEDS: Use a decoction as emetic; gastric problems; dyspepsia; hyper-activity; sciatica; diphtheria; intoxication poisoning.

PLANTAIN: Use as an infusion or fermentation for cuts, insect stings, bruises, and as an astringent. Pound fresh leaves, powder "kills worms of the belly,"... *A TREASURY OF AMERICAN HERBS,* by Virginia Scully: p. 225.

CATNIP: Decoction, infusion. Use ointment as a poultice for swelling. Tea for fevers, colds, headaches, menstrual pain.

YARROW: High in potassium. Use a decoction as a stimulant tea: one cup is enough. Use infusion for burns, blood cleansing, flu remedy, kidney disorders, bladder infections, antiseptic, mouthwash. Chew raw *leaves* for relief of toothache. Use ointment for skin problems.

BURDOCK SEEDS & BURRS: Use ointment for boils and skip eruptions. Ferment or infuse as blood purifier, diuretic. A source of insulin, chemical. Use to wash pets to kill fleas.

PIGWEED: High in calcium. Eat raw or make decoction. Use to compensate for lack of calcium in diet.

WILLOW, WEEPING OR WILD: Infusion or raw, an antipyretic. *Twigs* are a source of acetasalic acid or aspirin. Also found in the *bark*.

BLACKBERRY: Make the *bark* or *roots* into a syrup. Use as a tea or a constipation medication.

VIOLET LEAVES: As decoction or infusion. Use for heart pain, hyperacidity, asthma, rheumatism, headaches, tonsils, inflammation of the tongue, laryngitis, syphilis, blood cleansing tea. Use as an ointment for carbuncles.

BALSAM: Use *pitch*. Gather it by popping the blisters on the tree and letting the pitch run into a vial. Store and use as an instant glue for small cuts.

FALSE SOLOMON SEAL: Slice fresh *roots* and thread on a necklace; keep in a warm place. Powder as much as is needed and apply to cuts or wounds. Use a raw poultice of fresh *roots* for boils and sprains. The pulped, raw *root* has been used as an antiseptic applied to the eyes and ears. Use as an ointment for infected wounds.

TANSY: Make powder from button tops and use as an infusion for worms. Use as an ointment for relief of bruises, sprains, swellings. Use as a decoction to relieve superficial skin ulcers. Tansy has all of the stimulant qualities of ginger, but it is not edible as such.

HEAL-ALL: As a decoction, use as a diuretic, diaphoretic or tonic. As an ointment, relieves hemorrhoids.

STRAWBERRY LEAVES: Raw, source of iron and vitamin C, as a decoction, a tonic and a stimulant tea.

RED CLOVER: Raw, high in protein, vitamin A. Eat raw or powdered. Add as salt to the diet. As an infusion it reduces acid and intestinal gas, relieves bronchial cough, and acts as a blood tonic.

CHAMOMILE: High in calcium and vitamin K. As an infusion, a somatic, diuretic; use as a rinse for hair washing, or ointment for chapped skin.

THYME: Powder for use as an infusion or as a stimulant tea, intoxication remedy, vertigo, enteritis, hepatitis, rickets, gout, sciatica, neurosis, headache.

WILD LETTUCE: Pliny's Natural History lists 42 disorders. Milk mild sedative. Syrup of strong infusion induces sleep, calming effect. Relieves pain and induces sweat.

DANDELION: A source of calcium, vitamins A and K. Powder for infusion. Blood tonic.

GOLDENROD: Use as an infusion for gall stones, chlolithiasis; stops internal hemorrhages.

Notes

Upper left to right: Red clover, yarrow, dandelion, plantain, (short leaf, long leaf)
Lower left: White clover, strawberry leaves

Common lawn foods

TEAS:
Chamomile (tranquilizer)
Thyme (stimulant)
Rose Hips (vitamin C)
Mint (stomach soother)
Pine Needle (vitamin C – year round)

VEGETABLES...SALADS
(raw, steamed, frozen, canned)
Red Clover
White Clover
Dandelion
Plantain
Sorrel
Mustard
Violets
Chicory

MEDICINES:
All of the above vegetables plus
Heal-All.

VITAMIN C SOURCES: (besides the fruit below)
Balsam-year- round
Strawberry
Pine Needle-year-round
Rose Hip
Field Sorrel

FRUIT:
Blackberries-C source in dried berries, year-round, boil
Raspberries-C source in dried berries, year-round, boil
Rosehips-C source in dried hips, year-round, boil

The following grow best in pots, indoors. The conditions are the same as for your usual house plants. Same plants found on common lawns. Bring in to pots before fall to continue best growth.
Clover
Dandelion
Yarrow
Thyme
Mint
Sorrel (field, wood)
Mullein
Chicory
Plantain
Chamomile
Violets

HOT DANDELION SALAD

Ingredients:

4 handfuls of fresh Dandelions
Queen Anne's Lace salt
Thyme pepper

Wash tender new leaves. Soak in salted water to remove any bitterness, if desired. Place ham fat (or bacon drippings, butter, sesame or olive oil) in skillet. Stir in leaves briskly until wilted. Add thyme and sprinkle with Queen Anne's Lace. Serve HOT!

FRESH DANDELION SALAD

Ingredients:

4 cups fresh Dandelion greens
garlic
boiled egg (if desired)

Soak leaves in salted water if desired. (1-2 hours). Rub garlic into salad bowl. Place leaves in bowl. Add dressing (oil and vinegar or herbal vinegar) and egg slices. Serve.

DANDELION FLOWERS

Ingredients:

2 cups dandelion flowers
sesame or olive oil
garlic powder

Fry flowers in oil for 5-7 minutes. Sprinkle with garlic powder. Serve hot as garnish.

Fresh tender new shoots are the best, 1" to 3" ... you can find these under the snow by putting 4" to 6" of leaves over a dandelion area on your lawn. This acts as a greenhouse...boards can act the same. I have picked leaves under 4" of snow. The leaves were tiny and green under the old leaves in the snow. Hay works well also, 6-8" thick.

COMMON LAWN FOODS

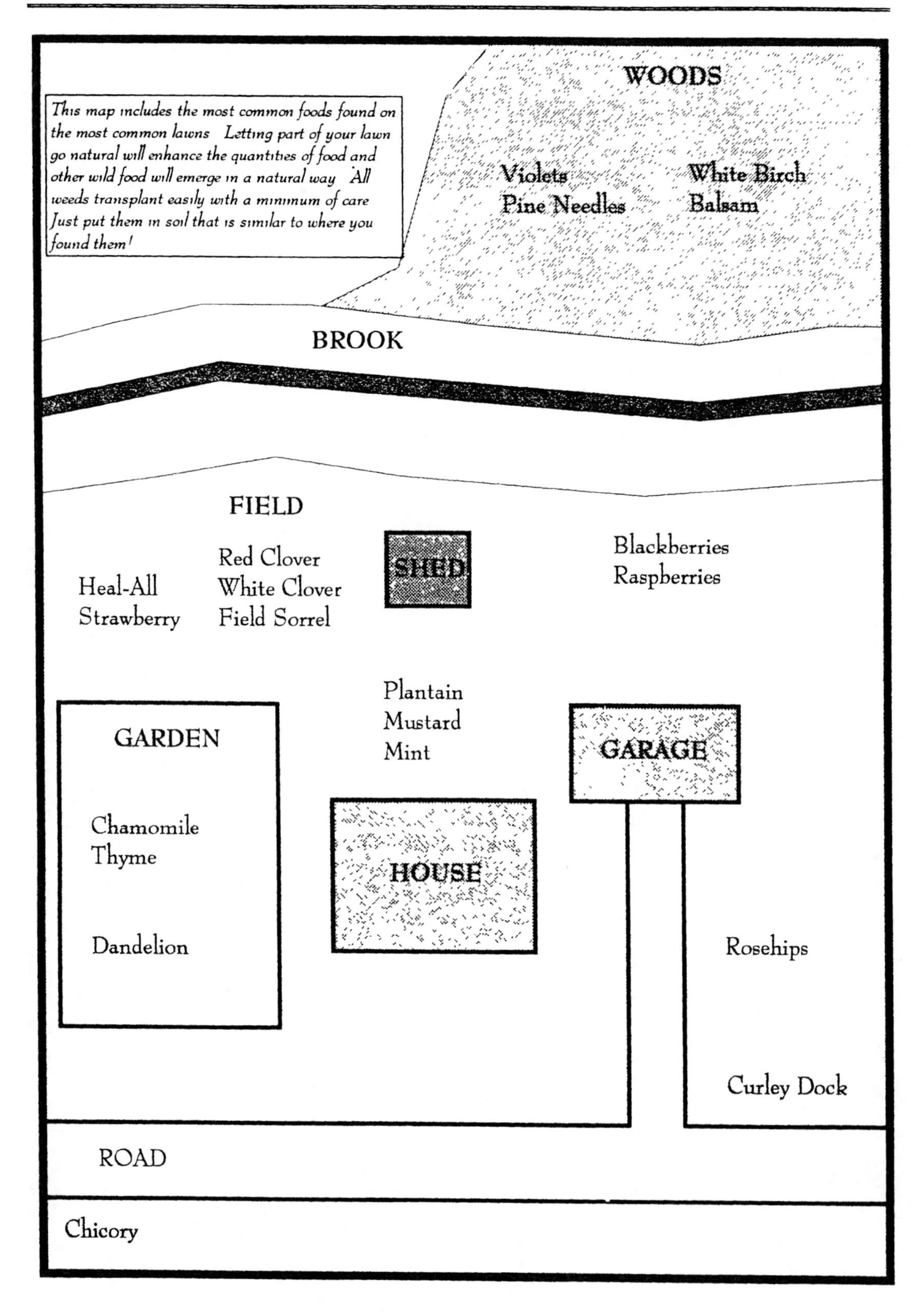

References

Common Poisonous Plants, John M. Kingsbury, Cornel University
The Dictionary of Useful Plants, Nelson Coon. 1974.
A Treasury of American Indian Herbs, Virginia Scully. 1970
The Book of Herbs, Edmond Bordeaux Szekely, 1978.
Herbal Remedies, Simmonite & Culpeper, 1970.
Eat the Weeds, Ben Charles Harris, 1973.
Are You Confused?, Paavo Airola, Ph.D., N.D. 1979.
Using Wild &. Wayside Plants, Nelson Coon, 1980.
Wild Foods, Jim Briggs, Cooperative Extension, Hamilton County.

NOTES